Terrific Toddlers

Time to Go!

by Carol Zeavin, MSEd, MEd
and Rhona Silverbush, JD

illustrated by Jon Davis

Magination Press • Washington, DC • American Psychological Association

Books for Kids From the
American Psychological Association

Magination Press is a registered trademark of the American Psychological Association. Order books at maginationpress.org,
or call 1-800-374-2721.

Book design by Gwen Grafft
Printed by Worzalla, Stevens Point, WI

Library of Congress Cataloging-in-Publication Data
Names: Zeavin, Carol, author. | Silverbush, Rhona, 1967- author. |
    Davis, Jon, 1969- illustrator.
Title: Time to go! / by Carol Zeavin, MSEd, MEd, and Rhona
    Silverbush, JD ; illustrated by Jon Davis.
Description: Washington, DC : Magination Press, [2020] |
    Series: Terrific toddlers | Summary: "When it's time to go,
    we aren't always ready to leave"— Provided by publisher.
Identifiers: LCCN 2019043249 | ISBN 9781433832529 (hardcover)
Subjects: LCSH: Railroads—Juvenile literature. | Travel—
    Juvenile literature.
Classification: LCC TF148 .Z43 2020 | DDC 625.2—dc23
LC record available at https://lccn.loc.gov/2019043249

Manufactured in the United States of America
10 9 8 7 6 5 4 3 2 1

With gratitude to my inspiring teachers and mentors
at Bank Street, Rockefeller, and Barnard—CZ

Dedicated to the inspiration for this series
(you know who you are!), with infinite love—RS

For Laura and Greta—JD

More Terrific Toddlers

Sometimes it's time to go.
But sometimes we don't want to go!

Ava is singing and banging on a pot.
Her daddy says, "It's almost time to go to the playground."
Ava says, "I singing!"
Daddy says, "OK, one more song, and then we go."

After one more song, Ava is still singing!
Daddy says, "I know you want to keep singing, but now it's time to go.
First we put on our shoes, then we take our snack.
Do you need help today?"

"No!" Ava says. "I do it myself!"
Daddy nods. "And then we say,
'Bye-bye, pot! See you next time!'"

Later, Ava, Jack, and JoJo
are playing in the sandbox.
"JoJo, time to go," her mommy says.
JoJo keeps shoveling.

Mommy says, "JoJo,
it's time to go home."
"NO!" JoJo yells.

She runs to the slide.

Mommy goes to her and says,
"I wish you could stay, but it's time to go.
You can slide down, or I will help you jump."

JoJo slides down. "Whee!"
Mommy catches her at the bottom.
"And now we say...

'Bye-bye, playground! See you next time!'"

When Jack gets home from the playground, he runs to his trains. Mommy says, "I'm getting your bath ready. You are full of sand!"

Jack pushes his trains around
and around the tracks. "Choo-choo!"
"Bath's ready!" Mommy says.
"I busy," Jack says.

Mommy says, "I see you're busy,
but now it's time to take your bath."
"No bath!" Jack says.

Mommy says, "Hmm. Do your trains need a bath, too?
You can push your trains to the tub. Choo-choo!"

Jack scoots his trains to the bathroom. "Choo-CHOO!

Bye-bye, tracks! See you next time!"

When it's time to go...we say,
"See you next time!"

Note to Parents and Caregivers

Our toddlers live in the moment. They like what they like, and they don't like change. They fight any and all limits. They hang onto control, because in reality they have so little of it.

Time to go — let the battle begin!

It's not funny, we know. These everyday frustrations can affect your loving relationship with your toddler. *Time To Go!* offers the following suggestions:

Give a head's up when possible. It's helpful to signal to toddlers that their current activity is coming to an end. Ava's dad uses "toddler-time" ("after one more song"), but you can also use a term like "five more minutes." As opposed to when you're leaving, when your toddler needs a more concrete sense of when you're returning (see *Bye-Bye!*), it's fine here to use a term like "five more minutes." They don't know how long five minutes is, so your five minutes can really be whatever length works for you... The important thing is giving the signal that it's almost time to switch gears.

Have a routine. Have a "first this, then that" routine, such as Ava's dad's "first shoes, then snack," which can guide them through the transition. Also, toddlers can express their need for control by learning to do parts of the routine themselves.

Use transitional objects. These are things they've been playing with and don't want to leave behind, like Jack's trains. Toddlers can carry them along to ease the transition to the next activity.

Be matter of fact. They're not really TRYING to drive you crazy...Be patient but firm — tell them you wish they could keep playing, because you do, for their sake. Label what's happening ("You really like the playground, but now it's time to go"), so they can understand themselves, and feel understood.

Give a choice... Offer two choices, as JoJo's mom does, both of which are okay with you. "Do you want to carry your shovel, or should I?" "You can walk, or I will carry you." They can choose, exercising their control, and you can achieve the transition you need.

...and then go. Endless negotiations will not get the job done, and toddlers feel more secure when a grownup actually takes charge (in a benevolent way, of course...). Take a breath, explain that, since they are having a hard time, you will help them — and then pick them up and go.

Battles will ensue, tantrums will happen, but if you stay clear and firm yet understanding, you will both have an easier time when it's time to go.

We love toddler pronunciation! And we know toddlers are not yet able to pronounce the complicated consonants in "bath" and "tracks" and "next." We just didn't want to annoy you with approximated spellings of most toddlers' best efforts ("No baff!" "Bye-bye twacks," "See you ness time"). So, don't worry if your toddler can't speak as well as Jack does — Jack can't, either!

Carol Zeavin holds master's degrees in education and special education from Bank Street College, and worked for eighteen years in homes and classrooms with toddlers. She was Head Teacher at both Rockefeller University's Child and Family Center and at the Barnard Toddler Development Center, and worked for Y.A.I. and Theracare. She is a professional violinist living in New York, NY.

Rhona Silverbush studied psychology and theater at Brandeis University and law at Boston College Law School. She represented refugees and has written and co-written several books, including a guide to acting Shakespeare. She currently coaches actors, writes, tutors, and consults for families of children and teens with learning differences and special needs. She lives in New York, NY.

Visit terrifictoddlersbookseries.com

🐦 @CarolRhona

📷 @TerrificToddlersBooks

Jon Davis is an award-winning illustrator of more than 80 books. He lives in England.

Visit jonsmind.com

🐦 @JonDavisIllust

📷 @JonDavisIllustration

Magination Press is the children's book imprint of the American Psychological Association. Through APA's publications, the association shares with the world mental health expertise and psychological knowledge. Magination Press books reach young readers and their parents and caregivers to make navigating life's challenges a little easier. It's the combined power of psychology and literature that makes a Magination Press book special.

Visit www.maginationpress.org

 @MaginationPress